# THE NATURE OF DENALI

## CONTENTS

# THE NATURE OF DENALI

A GUIDE TO THE ENTRANCE AREA TRAILS OF DENALI NATIONAL PARK AND PRESERVE

## WELCOME

Denali National Park and Preserve encompasses six million acres of breathtaking wilderness. Trails that invite you to explore this wilderness have been developed and maintained along the first three miles of the park road. If you wish an easy to moderate dayhike near the entrance area and want to learn more about what you are experiencing, this booklet is for you. From moose to mountains, this is your introduction to some of nature's finest creations.

## GETTING STARTED

Before you strap on your day pack and set off down the trail, take a moment to familiarize yourself with the hiking tips outlined on the back cover of this booklet. For more detailed guidelines, talk to a ranger at the Visitor Access Center and read the park newspaper, the *Denali Alpenglow*. Investigate these sources and be an informed hiker, one who has a safe and meaningful time in Denali and who does not interfere with the animals and plants that make their home here.

## THE TRAIL DESCRIPTIONS

This booklet describes the significance of the features found along four of the entrance area trails in Denali National Park. Each trail features a basic theme associated with the area's natural and cultural history.

**Morino Loop Trail** - human history
(1.3 miles round trip, 100 ft. elevation change)
**Taiga Loop Trail** - The subarctic forest ecosystem
(1.3 miles round trip, 150 ft. elevation change)
**Mount Healy Overlook Trail** - wildlife habitats
(4.0 miles round trip, 1,700 ft. elevation change)
**Horseshoe Lake Trail** - preserving parks for all time
(1.4 miles round trip, 200 ft. elevation change)

*Map of the Entrance Area Trails* Following these sections, you'll find a brief overview of the **Rock Creek Trail** and the **Roadside Trail** as well as checklists for plants and animals common in this area.

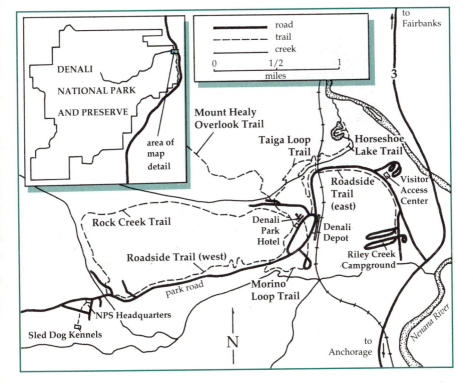

# MORINO LOOP TRAIL

## Human History

### A Changing Land Ethic

People have always valued this land, though the reasons may have changed over time. Walk the Morino Loop Trail to meet some of the characters from Denali's past and learn how their perceptions of this land have evolved through the years.

**Length:** 1.3 mile loop trail, 45 minutes to 1 hour total.

**Route:** The trail begins at the guest parking lot west of the Denali Park Hotel. Look for the Hiking Trails sign. The trail crosses the park road and passes through Morino Campground before returning to the hotel.

**Terrain:** 100 feet elevation change, half of the trail is gravel covered and half has an uneven surface.

**Highlights:** Black spruce bog, views of Hines Creek, and a panorama of the Alaska Railroad Trestle spanning Riley Creek.

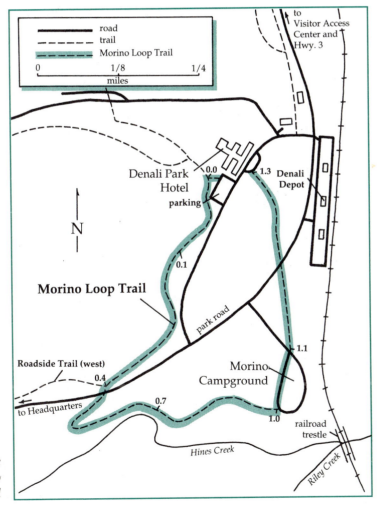

*Map of the Morino Loop Trail*

**START** Walk to the first junction 25 yards from the parking lot and turn left. The trail will move away from the road before reaching a rock cliff-face at mile 0.1.

• Athabaskan Survival

Whereas today we cherish this land for its majestic scenery, abundant wildlife, and recreation opportunities, the region's original inhabitants, the Athabaskan Indians, depended upon it for their survival. Ancestors of the original Athabaskans still live throughout the interior of Alaska. Denali means "the high one" in an Athabaskan language.

mile 0.1, rock cliff-face

Perhaps as long as 10,000 to 20,000 years ago and continuing until as recently as a few generations ago, these northern interior people made their living completely off the land. The archaeological evidence suggests that Athabaskans relied upon the Denali area as a hunting ground for caribou and other animals which provided food and materials for clothing and shelter. The rock promontory on the right of the trail about 0.1 mile from the parking lot is similar to those which were used as lookouts by Athabaskans waiting to ambush caribou as they migrated seasonally.

The trail continues through the spruce and aspen forest and at mile 0.4 passes the junction with the Roadside Trail and then crosses the park road. Look for traffic before crossing.

• Exploitation

mile 0.4, moose browse along the park road

As you reach the park road, look closely at the nicely pruned willow bushes that grow along the roadside. No, this is not the handiwork of park maintenance crews, but the tooth work of hungry moose. A large moose will eat up to 50 pounds of willow per day in order to maintain its body weight. In winter, the moose must resort to eating the branches of these deciduous trees and shrubs. Many aspens are stripped of their bark by hungry moose in late winter after other foods have been exhausted or covered with snow. Moose are just one of the game animals that, at the turn of the last century, attracted more people to this area.

Beginning in 1904, white settlers began prospecting for gold in the Kantishna area 65 miles west of here, as well as along other rivers in Alaska's interior. In response to this influx of people, market hunters began to make a regular business of shooting local game animals, selling the meat to nearby mining camps and Fairbanks. In the coming decade, market hunters slaughtered as many as 2,000 sheep each year. People began to exploit the land and its resources at a rate that could not be sustained for long.

**Beyond the road, the trail drops steeply for 100 yards before entering a dense forest and following along a level terrace above Hines Creek. Near mile 0.7, there is a view of the Alaska Railroad trestle crossing Riley Creek downstream.**

*Charles Sheldon*

• Conservation

In 1906, yet another hunter entered this area. Originally here to obtain museum specimens of Dall sheep, hunter-naturalist Charles Sheldon became concerned about the fate of local game animals. He recognized the value of resource conservation: limiting resource use and exploitation in order to ensure its continued existence. He realized that if Dall sheep and other animals were to survive, they would need a healthy habitat free of unchecked hunting pressures. In 1908, he returned to the lower 48 to share his concerns with politicians and members of the Boone and Crockett Club who were also making the shift toward a conservation ethic. With Sheldon's leadership, the club focused on a new idea - a park refuge that would preserve Denali's wildlife.

*mile 0.7, railroad trestle view*

The idea simmered for a few years until, in 1912, Alaska gained territorial status. Along with this event came plans for constructing a railroad through Alaska's interior. Conservationists were concerned about the additional pressure the market hunters would place on the game animals to supply railroad construction camps, towns along the rail line, and future travelers. Because of this threat, Sheldon and other politically influential citizens stepped up their efforts to protect Denali's wildlife. In 1917, the dream become a reality, the establishment of Mount McKinley National Park.

*mile 0.7, railroad trestle view*

The Alaska Railroad trestle comes into view just before the trail turns away from the creek and toward the park road. The railroad was built between 1915 and 1923. This trestle was the final link in the tracks between Seward, on the coast, and Fairbanks.

**The trail now climbs gently, passing through groves of aspen trees growing on the well-drained south-facing slopes.**

• Preservation

*mile 0.7 through 1.0, aspen forests*

Aspen, easily identified by their whitish-green bark and round, trembling leaves, are relative newcomers to this forest. Aspen need a lot of sun and got their start in this area after a fire burned the over-shading white spruce. After about 80 years, you can expect the white spruce, which get their start in the shade provided by the aspen, to once again replace the stands of aspen.

*1924 forest fire near the trestle*

The fire in 1924 was a human-caused fire that burned 30 square miles. At that time, the national park headquarters stood across the creek from here, near the south end of the rail road trestle. Harry Karstens, the park's first superintendent, and his staff of two rangers battled the

blaze for six days, fighting to keep the headquarters buildings from turning to ash. Karstens moved the headquarters buildings to their present location the following year.

Karstens' 1921 arrival in the fledgling national park began a new era of human perception of this land; preservation. These lands would forever be protected and preserved.

*Karstens in 1926*

This doesn't mean the area has not changed since the park was established. Just as the forest changes with time after a fire, parks go through changes that are both natural and human caused. Eventually, Mount McKinley National Park became accessible to even more people.

**After a short but steep climb, the trail reaches Morino Campground, a walk-in campground for people who travel to Denali without a vehicle.**

• Evolving Perceptions

*Morino's second roadhouse, 1940*

This campground and trail are named after Maurice Morino who homesteaded here in 1916, sixteen years before this land west of the Nenana River was added to the national park. Morino built a roadhouse near the north end of the railroad trestle to feed and accommodate railroad workers. In 1923, he built a second roadhouse to accommodate park visitors who travelled to the park via train.

**mile 1.0, Morino Campground**

During Morino's time, about 400 visitors made the journey to the park each year. For several decades, park visitation increased gradually, punctuated by a post war surge and the 1957 completion of the Denali Highway between Paxson and Cantwell. A dramatic change, however, occurred between 1971 and 1972 when the completion of the George Parks Highway linking Anchorage and Fairbanks doubled visitation. Today, the escalating annual Denali National Park visitation exceeds 500,000.

**mile 1.0, Morino Camp-ground** The features along this trail reflect not only the evolution of increasing visitation, but also the changing human perception of land. For thousands of years Athabaskans depended on the land for survival. Then came a brief period of resource exploitation, followed by the dawning of an age of resource conservation and, finally, the present stage of recreation and land preservation. One of the keys which has driven this process has been accessibility. The Athabaskans moved with the seasons, the railroad allowed easy access to the land, and eventually the year round highway conveniently attached Denali to the rest of Alaska and to the world.

As you return to the hotel, think about your travel to this wilderness park and how you perceive this land. Why do you value this land? How might our actions affect the qualities we value?

**FINISH** **Continue through the campground for 0.1 mile and take the trail to the park road. After crossing the park road, take the short (0.2 mile) trail back to the hotel.**

# TAIGA LOOP TRAIL

## The Subarctic Forest Ecosystem

### Simply Complex, the Taiga Forest Paradox

At first glance, Denali's forests may seem scrawny, monotonous, and disproportionate compared to the vast Alaskan landscape. The subarctic forest, or taiga, is sparse in terms of the number of different plants that can survive here, especially compared to forests farther south. Like all ecosystems, however, it has complex interconnections. Walk the Taiga Loop Trail, where the structural simplicity makes it easy to see and understand these interconnections.

**Length:** 1.3 mile loop trail, 45 minutes to 1 hour total.

**Route:** The trail begins near the guest parking lot west of the Denali Park Hotel. Look for the Hiking Trails sign. The trail loops behind the hotel, crosses Horseshoe Creek, and ends near the Mercantile just east of the hotel. This trail provides access to the Rock Creek Trailhead (at 100 yards) and the Mount Healy Overlook Trailhead (at mile 0.25) and can be combined with the Horseshoe Lake Trail (at mile 0.8).

**Terrain:** The total elevation change is 150 feet. The trail is mostly level with uneven footing. There are short but moderately steep sections into and out of the Horseshoe Creek drainage.

**Highlights:** Spruce and aspen forests, two Horseshoe Creek crossings, and views of Mount Fellows.

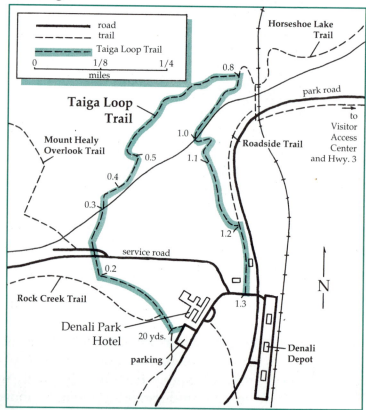

*Map of the Taiga Loop Trail*

**START**  Walk to the first junction just 20 yards up the trail. From here, the Taiga Loop Trail is to the right.

• The Living Parts

Look around and count the number of different types of trees you can see from this point. (You can use the checklist at the end of this guide booklet to keep track of the ones you can identify.) This area of Denali has just two species of cone bearing trees, white spruce and black spruce. The number of different types of broad-leafed trees can be counted on one hand. How does Denali's tree diversity, or number of different species, compare with your home? The taiga forest supports relatively few forms of life because few species can withstand the extreme growing conditions.

**20 yards, Taiga Loop Trailhead**

75 yards beyond the first junction, you'll pass the Rock Creek Trailhead. Our trail continues straight and then meanders through aspen groves for 0.2 miles before making an abrupt right turn near a rock outcrop.

• The Non-living Parts

So far you've probably focused your attention on the living parts of the taiga forest; the trees, shrubs, flowers, and animals. At this rock outcrop, take a moment to think about the non-living environment which includes things like temperature, nutrients, and moisture. The living and non-living parts of an area, along with all natural processes and interconnections, are collectively referred to as an ecosystem. In the next section of trail, we'll investigate how one environmental factor, climate, influences the survival of living things in this northern ecosystem.

**mile 0.2, rock outcrop**

The trail continues another 0.1 mile before reaching the first creek crossing. Along the way, you'll cross two service roads, the second of which provides access to the Healy Overlook Trailhead.

• Northern Climate

You might notice the air temperature getting cooler as the trail descends to and crosses Horseshoe Creek. Imagine how much colder it is in winter when the thermometer occasionally plummets to -50 degrees Fahrenheit! Lasting seven months, winter in Denali is the longest of all seasons. On the winter solstice, the shortest day of the year, Denali residents experience less than 4.5 hours of indirect light. How might that affect the growing conditions for plants and animals?

*mile 0.3, Horseshoe Creek crossing*

**The trail climbs gently as you leave the creek drainage. Fifty yards past the bridge, the trail turns abruptly to the right and then drops immediately into a dense forest of small evergreen trees.**

• Climate and Soil

*mile 0.4, black spruce forest*

Notice that there are very few aspens in this part of the spruce forest and that these cone bearing trees are smaller than those near the beginning of the trail. This is an area of black spruce, trees with small dark cones and a growth form that sometimes looks like a giant pipe cleaner. Black spruce usually grow in areas underlain by permafrost, ground that is permanently frozen due to the cold temperatures. Permafrost acts as an impenetrable layer so that in summer water pools on the surface creating bogs. Because water cannot move freely through the soil, acid builds up and nutrients are not always available to plants.

*black spruce growth form*

What other effects does the cold temperature have on the soil? In addition to freezing the ground, cold temperatures slow the processes creating fertile organic matter. Soil organisms which break down plant tissue into organic matter are less active in cool temperatures. Notice the thick layer of spongy vegetation, mostly moss, growing in this area. This insulates the soil and inhibits melting of the permafrost.

**Continuing on 0.1 mile, you'll move out of the permafrost zone.**

• Climate and Diversity

*mile 0.5, south facing slope*

Mount Fellows enters the southeast view across the valley as the trail rounds to a south facing slope where there is no permafrost. Once again, you're in an area of larger trees. Trees that can compete for these better sites are often the conical shaped white spruce. Although the white spruce do not have to deal with frozen ground, they too must withstand the extreme climate and short growing season. A white spruce that is 75 feet tall can be over 300 years old.

*white spruce growth form*

The subarctic climate not only determines the size of what can grow here, but it also creates pressures that limit the number of different species that can survive here. Denali's six million acres are home to about 750 species of plants. Six million acres in balmy Costa Rica, by comparison, can support over 9,000 different plant species.

**The trail meanders through the forest for the next 0.3 miles, first climbing and then dropping gradually.**

• The Forest and the Trees

*mile 0.5 through 0.8, a variety of trees*

Because there are few tree species in this forest, it doesn't take long to become an expert! In addition to the two evergreens, black spruce and white spruce, a few broad-leafed deciduous trees grow in Denali. There is a checklist of common trees and shrubs on page 42 of this guide.

Amble for the next 0.3 miles and look for:

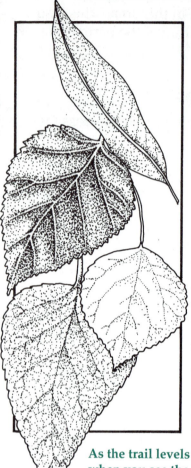

## Willow
Tall shrubs with long leaves that are often browsed by moose. They grow well along trails and streams. There are 24 species of willow in Denali so you're bound to find some variation.

## Paper birch
Trees that are easily identified by their white, peeling, paper-like bark. This bark is there for protection just like our skin. Please do not skin the trees!

## Aspen
Aspen have whitish-green, smooth bark and round, trembling leaves. They grow well on dry slopes.

## Balsam poplar
These are similar to aspen, but the bark of mature trees is often deeply furrowed and the leaves are arrow-head shaped. They grow well close to streams.

**As the trail levels out in an area of small black spruce, stop when you see the power line ahead.**

• Limitations and Adaptations

Look at this power pole and compare its size to the trees nearby. Do you think that this power pole tree grew near here? Trees in the interior of Alaska tend to be small because their growth is limited by the harsh climatic conditions.

**mile 0.8, power pole**

Many of the plants and animals that live here have built-in physical traits, adaptations, that give them an advantage in the extreme climatic conditions of the north. Small plants are better able to hug the ground where temperatures are usually warmer than the air temperature. Other adaptations to cold temperatures include being able to track the sun as it travels across the sky or having fuzzy or thick leaves for insulation. Look for shrubs nearby that have thick wax-covered leaves. Some of the common shrubs are lowbush cranberry, blueberry, and labrador tea.

*lowbush cranberry, blueberry, and labrador tea*

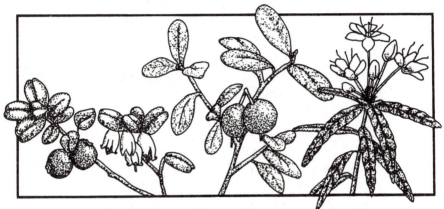

**At the junction with the trail to Horseshoe Lake Overlook, turn right. For the next 0.2 miles, the trail first drops steeply to and then follows along Horseshoe Creek.**

• Ecosystems

After you pass the junction with the spur trail to Horseshoe Lake Overlook, there is a striking vista to the south which must mean its time for . . . the big picture. Not only is this a good photo opportunity, but this is also an opportunity to think about a big concept - the taiga forest ecosystem. The taiga forest is an interconnected system of living and non-living parts and their natural processes. Ecosystems are not fully understood, but they are characterized by interdependent parts and change. We'll investigate these complex interconnections in the remaining 0.5 miles of the Taiga Loop Trail.

**mile 0.8 through 1.0, toward Horseshoe Creek**

**Stop for a moment at the bridge crossing Horseshoe Creek.**

• Interconnections

All things in this ecosystem are tied together and are dependent upon each other. For example, look for fireweed and other herbs growing along the trail. Plants are producers, able to convert sun energy, water, air, and nutrients from the soil into plant tissue by the process of photosynthesis. If you are watchful, you might see a consumer, a snowshoe hare, chewing the leaf of a fireweed, converting some of those nutrients into energy for its body. Perhaps a red fox will catch the hare and return to feed kits at its den. Eventually the foxes will grow up and die and the nutrients that have passed through so many lives will, with the help of bacteria and fungi, be released back into the soil. The nutrients will then be available again for plants, making the cycle complete. Each is an integral facet of a healthy ecosystem. Remove one and the health of the whole is compromised.

*mile 1.0, back across Horseshoe Creek*

**The trail climbs steeply away from the creek and levels out near a small cement building.**

• Change and Succession

It is easy to think of a forest as static and timeless, but ecosystems are actually changing all the time. After a disturbance opens an area to sun, the vegetation communities and the animals associated with them progressively change in a process called succession. In the taiga forest, it often begins with fire.

*mile 1.1, the pump shed*

A fire burned this area in 1924 which opened the forest to the sun's rays. Along this trail there is evidence of each phase of succession due to the mosaic pattern of burn in 1924 and more recent disturbances including road and trail building. Two to five years after a fire, fireweed, willow, and alder grow, improving the soil for future plants. Ten to 80 years after the fire, the shrubs are replaced by aspen in dry areas and by balsam poplar in wet areas. Young spruce grow in the shade provided by the aspen and eventually overtop the aspen in the final succession stage known as the climax stage. The climax forest is one that could potentially remain indefinitely, although that rarely happens in Alaska. Fire is a regular and natural part of this ecosystem.

**From the building, find the trail by keeping to the right. After meandering for 0.1 miles through a mixed forest and crossing two more old service roads, the trail meets the Roadside Trail near the park road.**

• An Intact Ecosystem

Currently, Denali's entire ecosystem is intact. That is, the populations of major predators are healthy, its natural processes are uninhibited, and it could potentially stay healthy forever. Denali, however, does not exist in a vacuum. There are other connections, **mile 1.2,** near and far, which affect its health. Nature does **junction** not recognize the political boundary we've drawn **with the** around this six million acre wilderness. The merlin, **Roadside** for example, travels to South America in winter. **Trail** There this small falcon may be subject to DDT poisoning and diminishing habitat. Similarly, Denali's boundaries cannot keep out noise and air pollution.

How do we fit into the system? People need to be aware of how our actions, here and at home, affect the integrity of Denali. We too are a living part of the cycle of life and are connected to the rest of the ultimate ecosystem, planet Earth.

**At the junction with the Roadside Trail, turn right and continue 0.1 mile to the Mercantile store. The trail ends within FINISH sight of the Denali Park Hotel.**

# MOUNT HEALY OVERLOOK TRAIL

## Wildlife Habitats

### Have to Have a Habitat

On this 1,700 foot climb, you'll pass through a variety of plant communities that provide homes for animals. As the altitude and other local conditions change, so do the animals. Animals are dependent upon these habitats for food, water, and shelter. Denali is a large, intact ecosystem consisting of a variety of habitats. It is valued world wide for its contribution to biological diversity.

**Length:** 2.0 miles one way, 3 to 4 hours round trip.

**Route:** Take the first 0.25 miles of the Taiga Loop Trail (see the description on page 13). At the second service road, turn left and walk about 50 yards to the Healy Overlook Trail sign. From here, it is 0.7 mile to the Scenic View and an additional 1.0 mile to the Mount Healy Overlook. Return by the same route.

**Terrain:** The trail gains a total of 1,700 feet in elevation. The first 1.0 mile to the Scenic View is moderately steep. The remaining 1.0 mile is very steep and strenuous.

**Highlights:** Spectacular views of the Denali National Park entrance area, the Nenana River valley, alpine ridges, and Healy Ridge.

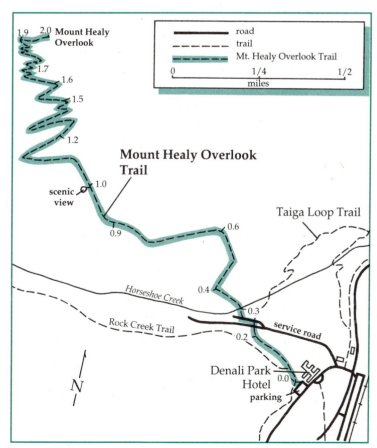

*Map of the Mount Healy Overlook Trail*

START Take the first 0.25 mile of the Taiga Loop Trail (read the description on page 13). At the second service road, turn left and walk 50 yards to the Healy Overlook Trail sign.

• Healy Ridge View

Look toward the high ridge to the northwest, Healy Ridge. Mount Healy Overlook is the rock outcrop you see on the horizon. This trail passes through forested lowlands and climbs above treeline to this promontory. The summit of Mount Healy is beyond.

mile 0.3, Healy Overlook Trailhead

There are two basic environments in this park and throughout much of interior Alaska. Taiga, a Russian word, is the term given to forested areas. Tundra, a Finnish word, is the term for areas without trees. There are, in addition, smaller environments such as streams, rock outcrops, thickets, and bogs, as well as transitions between all of these areas. As you walk along, you'll have the opportunity to see animals making their living. A place which provides food, water, and shelter is called a habitat. Please do not disturb the animals in their natural quest for food and protection. It's a matter of survival.

Stop at the bridge 20 yards ahead.

• Riparian Habitat

mile 0.3, Horseshoe Creek Bridge

Streams, in addition to providing a habitat for fish and aquatic organisms, have riparian areas along their stream sides. Riparian habitats provide moist environments for plants and animals. Willow, the tall shrub with narrow leaves, and balsam poplar, the trees with arrowhead shaped leaves and deeply furrowed bark, grow along this stream. Streams in Denali may also be home to mink or harlequin ducks. Ponds and lakes are home to beaver, muskrat, and many kinds of migratory birds that come to Alaska to take advantage of the seasonal bounty of insects. Many of these insects, including mosquitos, depend upon standing water for larvae development.

**Climb out of the stream bed and up a straight section of alder flanked trail for 0.1 mile.**

• Thicket Habitat

As the sound of water subsides, you'll enter yet another habitat where the willows give way to alder.

Thickets of alder flank the right side of the trail as it ascends. Alders are identified by their small cone-like fruiting structures and broad leaves with serrated edges. Spruce saplings grow in the shade provided by alders.

Notice that the alders grow mostly along the trail. Alders are pioneer plants that invade open areas soon after a disturbance such as trail or road building. Like legumes, alders have root nodules that contain bacteria able to change atmospheric nitrogen into a form plants can use. This usable nitrogen is returned to the soil when the leaves fall off and decay. The process is called nitrogen fixation which improves soil fertility.

*American green alder*

Snowshoe hares depend upon shrub thickets for protection. The hare is the principal food of many predators including lynx and red fox. The population of hares will climax and crash in roughly ten year cycles and will, in turn, affect the population of the lynx, which eats little else. Birds such as the arctic warbler and white-crowned sparrow also depend upon thicket habitats. The willow ptarmigan, the Alaska state bird, depends on these thickets at higher elevations.

*mile 0.3 through 0.4, an old road bed*

**As the trail bends to the right you'll enter an area where white spruce mix with aspen. Continuing on, the trail will bend left.**

## • Forest Habitat

Much of the taiga forest is made up of a single tree species, white spruce. White spruce grows in areas with moderately well-drained soils. Cow moose show a preference for this so called spruce moose forest when they have their calves in spring. Watch for moose, especially protective cows, and give them plenty of room. If they run at you, go ahead and follow your instincts and get out of the way quickly. Moose should calm down as soon as you get out of their space. Stay ate least 150 feet away if possible.

*mile 0.4 through 0.6, white spruce and aspen*

The white spruce provide food for many animals. Red squirrels spend much of their time in the fall collecting cones for a winter cache. Porcupines feast on the bark. Birds who inhabit this area include spruce grouse, boreal owl, boreal chickadee, three-toed woodpecker, and the very common gray jay, American robin, and dark-eyed junco. Where spruce occur with aspen and balsam poplar, look for Swainson's thrush, varied thrush, and white-winged crossbill.

*Left: black spruce Right: white spruce*

**The trail turns west, putting Healy Ridge on your right.**

## • Bog Habitat

Notice how the trees here are dramatically smaller and that the aspen are gone. Black spruce areas are usually underlain by permafrost. These areas are boggy in summer as the top layer of soil melts but has nowhere to drain. Weasels and martens may live nearby, feasting on voles that scurry through tiny tunnels beneath the dense moss. As the trail bends toward the ridge again, why are the trees even smaller but only in a swath leading straight up the ridge?

*mile 0.6 through 0.9, black spruce*

**Stop at the scenic view and rest on the benches.**

• Views and an Alder Swath

The scene at the viewpoint includes Mount Fellows to the east and the Alaska Range to the south. Also, notice how the area on either side of the trail is very different from the rest of the forest. As the trail ascends the ridge, very small spruce give way to a dense thicket of alder, the latter creating a swath heading straight up toward Healy Overlook. During World War II, Denali was turned into an off-duty recreational camp for Armed Forces troops. This trail zigzags through what was then a ski slope!

**mile 1.0, Scenic View**

Look to the side of the swath to observe the natural transition from taiga to tundra as your eyes move up the slope. Trees give way to shrubs, the transition zone, and finally, just out of view beyond the horizon, to tundra, the land above the trees.

From here to the Overlook, a distance of 1.0 mile, the trail climbs about 1,300 feet in elevation. Motivational hint: Many animals prefer the transition zone. Keep hiking!

**The trail climbs straight up the alder slope before turning to the left. After climbing into an area with spruce and paper birch, the trail turns to the right and continues straight. Stop at a large boulder for a breather.**

• A Glacial Landscape

After the trail switches back through the alder slope with occasional views, a welcomed rest stop eventually appears. The trail will take you behind and on top of a 12 foot high rock. Take the opportunity to catch your breath and look out over the expansive view.

**mile 1.2, halfway rock**

Glaciers are one of the primary forces creating this sweeping landscape of diverse habitats. The broad valley of the Nenana River to the left (southeast), the u-shaped valley of Riley Creek directly ahead (south), and the misfit boulders on the ridge farther to the right (southwest) are all clues which tell of a fairly recent ice age. As recently as 10,000 years ago, glaciers flowed from the Alaska Range. The Outer Range, which you are standing on, escaped the ice and has steeper v-shaped valleys carved by water. We are in an interglacial period; glaciers will eventually return to these valleys and change them again.

*The trail continues east and then west in a series of long switchbacks. The trail then creates a series of very short zigzags.*

• The Edge-Effect

The edges between thickets and clearings provide the best of all worlds for wildlife. Animals that typically derive shelter in a forest or thicket, such as the black-billed magpie whose streaming tail is longer than its body, will venture out into clearings to graze on the herbs, grasses, berries, and insects. For us, the clearings provide inspirational views. Please resist the urge to take shortcuts between switchbacks. This causes unsightly scars that begin the process of soil erosion.

*mile 1.2 through 1.5, scattered clearings*

*The trail continues to climb!*

• The Transition Zone

Eventually you'll leave most of the alder behind and come out into a rocky area below the Overlook. This transition from taiga to tundra, with just a few small spruce and alder scattered widely, attracts herbivores and the carnivores that hunt them. Many animals will find shelter in one area and feed in the other. For example, birds of prey such as the northern goshawk and sharp-shinned hawk build stick nests in forest trees and hunt for small mammals and birds in the transition area.

*mile 1.6, sparse spruce*

*The trail switches back below a rock outcrop.*

• Rock Habitats

Listen and watch for pika and marmot who make their homes among rocky slopes. The small plump-bodied pika, a close relative of the rabbit, might sound a nasal chirp and then duck quickly into a crevice. The larger hoary marmot, northern cousin of the ground hog, might sound a long, loud piercing whistle alarm as you approach its burrow or den. Also, look for birds such as the northern wheatear and Townsend's solitaire near rocky ridges in the park.

*mile 1.7, rock outcrop*

After climbing through several very steep, short, alder flanked switchbacks, the trail finally emerges immediately below the Healy Overlook. Continue to climb to its west facing slope.

• Tundra Habitat

You've climbed beyond the transition and onto the tundra. This dry tundra is made up of a thin mat of vegetation only two to three inches thick over rocky soil. Exposed to low temperatures and high winds, tundra plants are small in order to take advantage of warmer micro-climates close to the ground and behind rocks. Dry tundra provides grass and mountain avens for Dall sheep. Moist tundra, usually at lower elevations, supplies lichen to grazing caribou. Birds, including the lesser golden-plover, long-tailed jaeger, and Lapland longspur, migrate from nearly every continent to nest on the flourishing summer tundra.

*mile 1.9, tiny plants hugging the ground*

Tundra is home to the arctic ground squirrel whose body temperature can drop below freezing when hibernating. In summer and fall, these quick rodents make occasional meals for lucky grizzly bears. Typical grizzly fare consists of roots, grasses, and berries.

You deserve a rest. Sit right where you are and gaze north-west toward the summit of Mount Healy.

• Mountain Top Habitat

Dall sheep are often seen as white specks near the tops of rocky mountains. The concave soft-bottomed feet of Dall sheep allow them to scamper around on the steepest of slopes and this agility is their main defense against predators. They are much more vulnerable to predation when grazing down low or when making seasonal migrations to and from the windswept slopes of the Alaska Range.

*mile 1.9, Mount Healy Ridge*

Soaring hawks and golden eagles often catch the uplifting thermals above high tundra and mountain tops. These raptors, as well as swift peregrine and gyrfalcons, build large stick nests on cliffs and scan the landscape for meals.

**The trail continues through the tundra and emerges at last onto the rocky shoulder of Healy Ridge called the Healy Overlook. The actual summit of Mount Healy is far beyond.**

• Biosphere Reserve

Scientists realize that conservation is not accomplished by protecting single species, but by preserving the full array of habitats where species live. Intact ecosystems are valuable because they protect the diversity of life, or biodiversity, and provide a standard against which peoples' effects on the environment can be measured.

*mile 2.0, Mount Healy Overlook*

The United Nations Education, Scientific, and Cultral Organization (UNESCO) recognizes internationally significant examples of Earth's natural regions. In 1982, UNESCO recognized the value of this ecosystem in contributing to the preservation of biodiversity worldwide by designating Denali National Park and Preserve as an International Biosphere Reserve. The significance and value of Denali's habitats extend well beyond its boundaries.

**FINISH  Return by the same route.**

# HORSESHOE LAKE TRAIL

## Preserving Parks for All Time

### The Best Idea America Ever Had

Denali National Park and Preserve is a crown jewel of the National Park System and is one of the few remaining intact ecosystems on Earth. Where else does the synthesis of expansive wilderness, healthy wildlife populations, and natural processes create such a unique and dynamic landscape? The Horseshoe Lake Trail provides an opportunity to investigate some of Denali's natural wonders as well as our relationship to them.

**Length:** 0.7 miles one way, 1.5 hours round trip.

**Route:** The trail begins at mile 0.9 on the park road, near the small parking area at the junction of the park road and the railroad tracks. (You can walk to the trailhead on the Roadside Trail. It is 0.5 miles west of the Visitor Access Center and 0.3 miles east of the Denali Park Hotel.) The trail ends at Horseshoe Lake and returns by the same route.

**Terrain:** The trail descends 200 feet in elevation. The trail to the Horseshoe Lake Overlook (mile 0.2) is mostly flat. Beyond the Overlook, the trail drops very steeply. The return trip is moderately strenuous.

**Highlights:** Views of the Nenana River, an oxbow lake, and a beaver dam and lodge. Watch for abundant wildlife along this trail and at the lake.

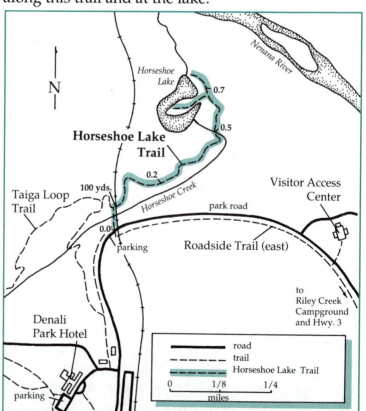

*Map of the Horseshoe Lake Trail*

**START**    Keeping alert for traffic, be it autos, trains, or planes, walk to the Horseshoe Lake sign near where the park road crosses the railroad tracks.

• The Expanding Mind Trick

**mile 0.0, Horseshoe Lake Parking Area**

Stand for a moment and picture where you are. Immediately surrounding you are signs of human development: the road, the air-strip, and the railroad tracks. But if you expand your perception a bit, you can picture yourself on the eastern edge of a wilderness the size of Massachusetts. Beyond that, let your mind picture the undeveloped expanse in all directions, from the Bering Sea to the east coast of Canada, and from the icy Arctic Ocean to the distant civilization of the "lower 48" where most Americans have their homes and daily lives. Where else can we feel so dramatically the notions of distance and space?

From the parking area, walk north along the railroad tracks 100 yards. Look for the trailhead sign and turn right onto the trail.

• The Great Attraction

**100 yards, Horseshoe Lake Trailhead**

National parks have been called the best idea America ever had. What is it about natural areas that attract people? Perhaps as you continue you'll be able to better answer that question. As you follow the trail to the lake, the human-caused sights and sounds that surround you will be gradually replaced by the signs of nature. You'll enter a realm where the body can relax, the mind can expand, and the spirit can be restored.

The trail undulates gently for 0.2 miles before reaching the Horseshoe Lake Overlook where there is a view of the lake and the Nenana River.

• The Paradox

At the Horseshoe Lake Overlook you can get a feel for the lay of the land. Below, there is an island-like peninsula in the center of the lake. Similarly, the area you are exploring is an island of nature in a sea of development.

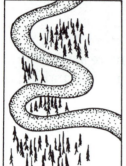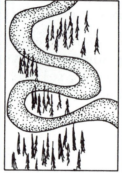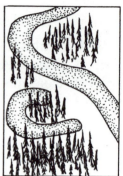

*Formation of an ox-bow lake*

Horseshoe Lake is an "ox-bow" lake, once a tight bend in the Nenana River that was cut off when the river eroded its more direct present course. The abandoned channel became the curved lake we see today. The lake basin is flanked by the railroad track to the west, the park road to the south, and the developed area along the George Parks Highway beyond the Nenana River to the east.

**mile 0.2, Horseshoe Lake Overlook**

How can two seemingly opposing forces, development and nature, be reconciled in a national park? This same paradox is embedded in the very essence of the National Park Service mandate. The Organic Act which created the National Park Service in 1916 states that parks should provide for visitor enjoyment but also conserve the scenery, wildlife, and other features in essentially their natural state. Park managers constantly grapple with achieving a balance between use and preservation. Developments necessary for our use and enjoyment include roads and accommodations, but these developments are built at the expense of wildlife habitat and natural scenery. How would you achieve a balance?

**The trail drops very steeply, with occasional flat areas, for the next 0.3 miles.**

• Animals and Access

As you head down the steep trail you can see one of the main Park Service developments, the Visitor Access Center (VAC) one half mile to the east. Most parks have visitor centers. Why is this one called an access center? In a park where wildlife is one of the key features, access to the park becomes a key con-

**mile 0.2** cern. Research indicates that unlimited access in the **through 0.5,** backcountry and along the Denali Park Road could **the steep** reduce the chance of seeing wildlife. So, although **trail down** parks are managed for people's enjoyment, numbers of people and the manner in which they enter the park play a role in determining the quality of the experience. A quick look around the VAC will show how park managers are helping to maintain a balance. Access to the park is restricted to shuttle busses, campground registration, and backcountry permits.

As you continue down the trail, the human developments gradually go out of view and are replaced by the signs of wildlife. Some animals apparently have become accustomed to people. Listen for the scolding chatter of red squirrels as they dart among the branches of the spruce trees. Keep a sharp lookout for signs of both snowshoe hare and moose, both of which eat willow and aspen bark as a winter food source. If the bark is torn off, this is probably the work of the moose. If it has been chewed off in small, uniform bites that begin a couple of feet off of the ground, this is a sign of snowshoe hare activity. Imagine the snowshoe hare as it chews its meal, its white winter coat nearly invisible against the snow pack.

**Eventually the trail will become level and will cross an intermittent stream bed.**

• Animal Signs

Notice how the spruce forest diminishes and water-loving balsam poplar take over as you approach a stream bed. The trail crosses the stream via a wooden bridge. Except for the trails in this entrance area, Denali has no developed trails and, consequently, no bridges. Backpackers, in a true wilderness tradition, must find their own route and wade across rivers. Denali's backcountry is wilderness in more ways than one. It is a legally defined wilderness, free of mechanized vehicle use and signs of human development. It is an environmental wilderness where natural systems prevail. It is also wilderness of the mind, where people can go to enjoy nature on its own terms. What does wilderness mean to you?

*mile 0.5, bridge over Horseshoe Creek*

As you continue to the lake, look for signs of beaver activity. Beavers are known for their ability to alter

*beaver signs*

their environment in order to secure shelter and survival. Look for stumps and fallen trees with characteristic teeth marks. The teeth of the beaver grow throughout their lives so that they can chew wood continuously. What do they do with their logs? Some of the leaves and inner bark, or cambium, become food, but most of the wood is transported to the lake for use as building materials. Continue down the trail to see some of their engineering handiwork.

**The trail meanders through the spruce forest for 0.2 miles and emerges along the eastern edge of Horseshoe Lake. From here, you can continue along the trail to the beaver dam at the far end of the lake. You can also cross to the peninsula in the center of the lake to view the beaver lodge on the north side of the "island".**

• Beaver Buildings

Please do not disturb the beavers or stand on the structures they have built! Both the dam and the lodge are vital to their survival and require constant upkeep. Please do not make the beaver's job even more difficult.

### The Dam

*mile 0.7, lakeside view*

When beavers built this dam, the level of the lake rose several feet. The dam creates two advantages, it expands the beavers' habitat and it prevents the lake water from completely freezing in winter. Sometimes beavers build multiple dams in order to further expand the space in which they can safely find trees and shrubs for food and shelter. Young beavers must move downstream to an unoccupied habitat in order to ensure that the family does not run out of trees.

*beaver lodge and dam*

### The Lodge

The beavers used small and large logs and packed mud for lodge construction materials. The beaver enter this lodge by a tunnel under water deep enough so that it does not freeze in winter. The tunnel leads to a room which is above the water surface. In fall, beavers cache aspen and willow branches near the underwater entryway so they can easily reach the food under the ice in winter, thereby escaping predators such as the wolf.

Like beavers, people are known for their ability to manipulate their environment and build structures. In a national park, some developments like roads, trails, hotels, and visitor centers are necessary so that people can access the park and safely enjoy themselves. However, these developments sometimes clash with the broader principle that a national park is a place where natural features and natural processes have precedence. It is a difficult balancing act - use vs. preservation. The key is outlined in the 1916 Organic Act which states that we can enjoy our parks, but only in such a way as will leave them "unimpaired for future generations." Denali is a good place to begin. If we can't strike the balance here, then where?

**FINISH  Return by the same route.**

## OTHER TRAILS

The trails described on the following pages offer further opportunities for adventure in the entrance area of Denali National Park. These trails do not have lengthy descriptions, but basic information is provided. Choose a trail that fits your interests and abilities.

## COMBINING YOUR HIKE WITH A SLED DOG DEMONSTRATION

For variety, combine either the Rock Creek Trail (2.3 miles one way) or the Roadside Trail (1.8 miles one way) with a Sled Dog Demonstration, a program presented several times daily in summer at the Dog Kennel near NPS Headquarters. Inquire at the Visitor Access Center for times and details.

- **Hiking to the Demonstration (uphill)**

If you wish to walk to the demonstration from the Denali Park Hotel, read the descriptions on the following pages.

- **Hiking to the Hotel After the Demonstration (downhill)**

If you prefer, catch a free shuttle bus to the demonstration and walk back to the hotel. The shuttle bus boards at either the Denali Park Hotel or the Visitor Access Center. After the demonstration, follow the signs to the visitor parking area near the park road. Walk east (to the right) on the park road for about 300 yards and cross the Rock Creek Bridge. Look for the Rock Creek Trailhead on the north side of the road immediately east of the bridge. To get to the Roadside Trailhead, continue east on the park road for 50 yards and look for the trailhead immediately after crossing a service road.

# ROCK CREEK TRAIL

**Length:** 2.3 miles one way, 2 hours one way.

**Route:** The Rock Creek Trail begins at the parking lot west of the Denali Park Hotel. Take the Taiga Loop Trail for 100 yards and then look for the Rock Creek Trailhead on the left. From this junction, the trail climbs steeply at first and then more gently along the Horseshoe Creek drainage. The trail then undulates through a spruce and aspen forest before dropping to Rock Creek. The trail ends near the Rock Creek Bridge at mile 3.0 on the park road. You can make a 4.1 mile loop trip by combining this route with the Roadside Trail described on the following page.

*Map of the Rock Creek Trail and Roadside Trail (west)*

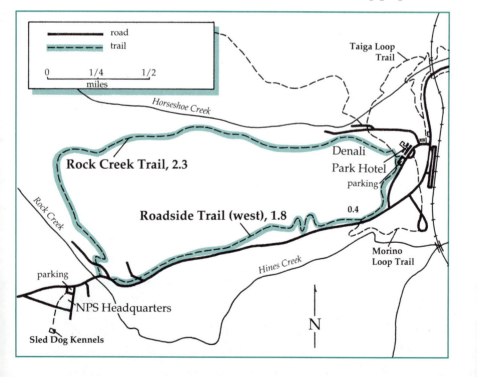

road

trail

0    1/4    1/2
miles

Horseshoe Creek

Taiga Loop Trail

**Rock Creek Trail, 2.3**

Denali Park Hotel

parking

Rock Creek

**Roadside Trail (west), 1.8**

0.4

parking

Hines Creek

Morino Loop Trail

NPS Headquarters

Sled Dog Kennels

N

Terrain:  The trail climbs a total of 400 feet.  The grade is very steep for the first 0.5 miles but then becomes more gentle.  Near Rock Creek at the end of the trail, the trail drops steeply before crossing a service road and reaching the park road.

Highlights:  Along the trail there are striking views of Horseshoe Creek and Healy Ridge.  This trail takes you far beyond the noise and congestion surrounding the hotel and the park road.

# ROADSIDE TRAIL (west)

**Linking the Denali Park Hotel with NPS Headquarters**

Length:  1.8 miles, 45 minutes to 1 hour one way.

Route:  Start at the parking lot west of the Denali National Park Hotel.  Look for the Hiking Trails sign.  Take the Morino Loop Trail for 0.4 miles and, just before you cross the park road for the first time, look for the Roadside Trail Junction on the right.  The Roadside Trail parallels the park road for most of its length except for 0.5 miles when it moves away from the road and switches back up a slope.  The trail ends near the service road just east of the Rock Creek Bridge at mile 3.0 on the park road.  You can make a 4.1 mile loop trip by combining this route with the Rock Creek Trail described on the previous page.

Terrain:  The trail climbs 300 feet in elevation.  There is a 0.5 mile section of moderately steep grades.  The rest is fairly level.

Highlights:  This trail features aspen, paper birch, and spruce forests.

# ROADSIDE TRAIL (east)

**Linking Riley Creek Campground
with the Denali Park Hotel**

**Length:** 1.2 miles, 45 minutes one way.

**Route:** This eastern section of the Roadside Trail follows immediately along the park road and connects Riley Creek Campground with the Visitor Access Center, the Horseshoe Lake Trailhead, and the Denali Park Hotel area.

**Terrain:** This trail gradually gains 100 feet in elevation.

**Highlights:** Hiking this trail is safer than walking the park road. This part of the park road was resurfaced in 1990, creating some resource damage along the roadside. The following year, resource managers assisted natural plant colonization in this highly visible area. Most of the vegetation growing between the trail and the road was grown from seeds native to Denali and was planted here by hand.

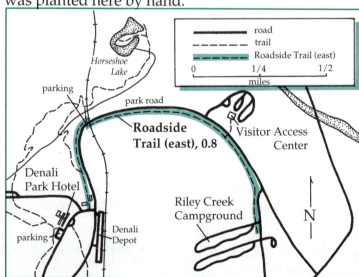

*Map of the
Roadside
Trail (east)*

# ANIMALS COMMON IN THE TAIGA FOREST

## MAMMALS

Insectivores

___ shrew (several species)

Bats

___ little brown bat

Hares

___ snowshoe hare

Rodents

___ arctic ground squirrel
___ red squirrel
___ northern flying squirrel
___ beaver
___ vole (several species)
___ muskrat
___ lemming
___ porcupine

Carnivores

___ coyote
___ wolf
___ red fox
___ black bear
___ grizzly bear
___ marten
___ weasel
___ wolverine
___ lynx

Cloven-hooved Mammals

___ moose

## BIRDS

Hawks and Falcons

___ goshawk
___ merlin
___ American kestrel

Grouse

___ spruce grouse

Owls

___ great-horned owl
___ hawk owl

Woodpeckers

___ common flicker

Flycatchers

___ olive-sided flycatcher

Corvids

___ gray jay
___ black-billed magpie
___ common raven

Chickadees

___ black-capped chickadee
___ boreal chickadee

Thrushes

___ American robin
___ varied thrush
___ Swainson's thrush
___ gray-cheeked thrush

Kinglets

___ arctic warbler
___ ruby-crowned kinglet

Wood Warblers

___ orange-crowned warbler
___ yellow warbler
___ yellow-rumped warbler
___ Wilson's warbler

Finches and Sparrows

___ common redpole
___ white-winged crossbill
___ dark-eyed junco
___ tree sparrow
___ white-crowned sparrow
___ fox sparrow

# PLANTS COMMON IN THE TAIGA FOREST

## TREES AND SHRUBS

Pine Family
___ white spruce
___ black spruce
___ common juniper
Willow Family
___ willow (various species)
___ quaking aspen
___ balsam poplar
Birch Family
___ dwarf birch
___ Alaska paper birch
___ American green alder
Saxifrage Family
___ northern red currant
Rose Family
___ shrubby cinquefoil
___ prickly rose
Oleaster Family
___ soapberry
Crowberry Family
___ crowberry
Heath Family
___ labrador tea
___ kinnikinnik
___ red bearberry
___ lowbush cranberry
___ blueberry
Honeysuckle Family
___ highbush cranberry

## FLOWERING PLANTS

Lily Family
___ death camas
Orchid Family
___ northern green orchid
Sandalwood Family
___ bastard toadflax
Crowfoot Family
___ pasque flower
___ larkspur
___ monkshood
Saxifrage Family
___ grass-of-Parnassus
___ three-toothed saxifrage
Pea Family
___ milk vetch
___ lupine
___ pea vine
Evening Primrose Family
___ tall fireweed
Dogwood Family
___ dwarf dogwood
Wintergreen Family
___ shy maiden
___ large-flowered wintergreen
Gentian Family
___ four-parted gentian
Phlox Family
___ tall Jacob's ladder
Borage Family
___ bluebells
Honeysuckle Family
___ twinflower
Valarian Family
___ valarian
Composite Family
___ yarrow
___ arnica
___ Siberian aster
___ coltsfoot
___ black-tipped groundsel
___ goldenrod

# FIELD
# NOTES
# AND SKETCHES

# FIELD
# NOTES
# AND SKETCHES

# FIELD
# NOTES
# AND SKETCHES

# FIELD
# NOTES
# AND SKETCHES

# FIELD
# NOTES
# AND SKETCHES

# FIELD
# NOTES
# AND SKETCHES